19.92

INTERIM SITE

19.92

INTERIM SITE

FACING THE FUTURE

SPACE SCIENCE

...oduction

...ence was born at a time lost in history
...ople first looked up at the stars and
...ed what they were. During the thou-
...f years between then and now, science
...now it has evolved, and scientists have
...rmed those early wonderings into pre-
...eories about the stars, the planets, and
...ace between them.

...il the space age, theories about space
...only be tested by observations and
...urements made from Earth's surface. Un-
...nately, Earth's **atmosphere** either blocks
...istorts almost all of the useful **radiation**
...reaches Earth from distant parts of the
...verse. For example, stars appear to twinkle
...an observer on Earth because light rays are
...isted as they pass down through the atmos-
...here. This prevents astronomers from pro-
...ucing clear images of very distant stars and
...alaxies. However, the space age has given
...cientists the opportunity to place satellites
...utside Earth's atmosphere, where they have
a much clearer view of the universe.

The space age began on October 4, 1957
when the former Soviet Union launched a tiny
29-inch spherical satellite called *Sputnik 1.* It
was the first object to be sent beyond Earth's
atmosphere and placed in space. Since then,
thousands of satellites have been launched,
mainly by the United States and the former
Soviet Union. They are no longer tiny spheres.
The largest weigh several tons and their electronic **payloads** are much more advanced than
the simple radio transmitter on *Sputnik 1.*

Earth Resources Satellites

Satellites are increasingly being used to explore Earth from space. Huge areas of Earth
have not been surveyed in detail from the
ground. Satellite images enable these areas to
be mapped and surveyed relatively cheaply.

Different types of soil, rocks, and vegetation
reflect the sun's energy in different ways. Satellites carry **sensors** that can detect these differ-
ences, which show up as different colors in the
satellite image. Using this method, new copper
ores have been found in Bolivia and Colombia,

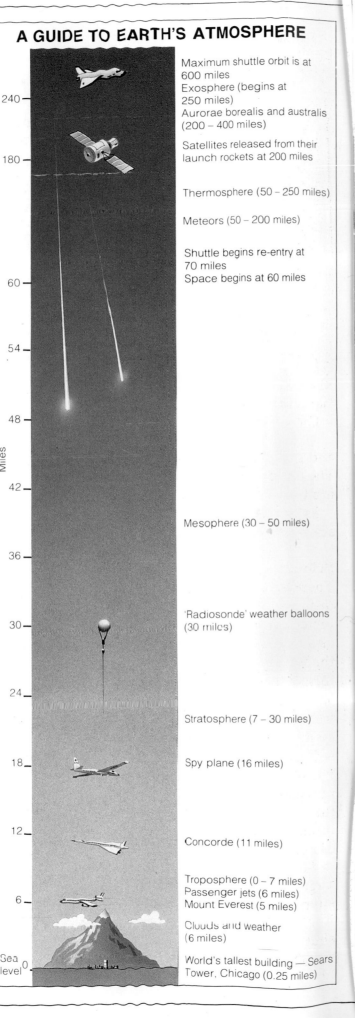

A GUIDE TO EARTH'S ATMOSPHERE

Maximum shuttle orbit is at 600 miles
Exosphere (begins at 250 miles)
Aurorae borealis and australis (200 – 400 miles)

Satellites released from their launch rockets at 200 miles

Thermosphere (50 – 250 miles)

Meteors (50 – 200 miles)

Shuttle begins re-entry at 70 miles
Space begins at 60 miles

Mesophere (30 – 50 miles)

'Radiosonde' weather balloons (30 miles)

Stratosphere (7 – 30 miles)

Spy plane (16 miles)

Concorde (11 miles)

Troposphere (0 – 7 miles)
Passenger jets (6 miles)
Mount Everest (5 miles)

Clouds and weather (6 miles)

World's tallest building — Sears Tower, Chicago (0.25 miles)

Miles — 240, 180, 60, 54, 48, 42, 36, 30, 24, 18, 12, 6, Sea level 0

FACING THE FUTURE

SPACE SCIENCE

Ian Graham

RSVP
RAINTREE
STECK-VAUGHN
P U B L I S H E R S
The Steck-Vaughn Company

Austin, Texas

Library of Congress Cataloging-in-Publication Data

Graham, Ian, 1953–
     Space science / Ian Graham.
         p.   cm.—(Facing the future)
     Includes index.
     Summary: Describes the various ways in which we have examined and explored outer space and possible future ways of using this knowledge.
     ISBN 0-8114-2806-0
     1. Space sciences—Juvenile literature.   [1. Outer space— Exploration.   2. Space sciences.]   I. Title.   II. Series.
QB500.22.G73   1993
500.5—dc20   92-18319
                                                              CIP
                                                              AC

Typeset by Tom Fenton Studio, Neptune, NJ
Printed in Hong Kong
Bound in the United States
1 2 3 4 5 6 7 8 9 0 HK 98 97 96 95 94 93

Acknowledgments

Illustrations – Graham White, Graham-Cameron Illustration
Design – Neil Sayer
Series editor – Su Swallow
Editor – Nicola Barber

For permission to reproduce copyright material the author and publishers gratefully acknowledge the following:

Cover photograph – artist impression of solar satellite power system (a possible future energy source) under construction. Boeing Aircraft Co., Science Photo Library
Title page – the Hermes spaceplane – European Space Agency

Page 5 – Tom Van Sant/Geosphere Project Santa Monica, Science Photo Library; page 6 – European Space Agency; page 7 – from an illustration that first appeared in *New Scientist*, London, the weekly review of science and technology; page 8 (left and right) European Space Agency, Science Photo Library; page 9 – Earth Observation Satellite Co. (EOSAT) Lanham Maryland USA – (inset) Bill Wood, Bruce Coleman Limited; page 10 – (left) Sinclair Stammers, Science Photo Library – (right) Earth Satellite Corporation, Science Photo Library; page 11 – J.D. Griggs, G.S.F. Picture Library; page 12 – (left top and bottom) NASA – (middle) Alex Bartel, Science Photo Library – (right) Nigel Cattlin, Holt Studios Ltd; page 13 – Dr. Gene Feldman/NASA GSFC, Science Photo Library; page 14 – NASA; page 15 – Science Photo Library; page 16 – Hughes Aircraft Company; page 17 – (left) Ford Motor Company – (right) NASA, Science Photo Library; page 18 – Robert Harding Picture Library – (inset) Navastar Ltd; page 21 – from an illustration by Gary Cook, *The Sunday Times*, 25 February 1990; pages 22-23 – Martin Marietta; page 24 – Hughes Aircraft Company; page 25 – NASA, Science Photo Library; page 26 – Jack Finch, Science Photo Library; page 27 – (top) Hencoup Enterprises, Science Photo Library – (inset) Roger Ressmeyer/Starlight, Science Photo Library – (bottom) Space Telescope Science Institute/NASA, Science Photo Library; page 28 – (left) Dr. Leon Golub, Science Photo Library – (top) NASA/Science Photo Library – (bottom) Robert Harding Picture Library; page 29 – NASA, Robert Harding Picture Library; page 30 – (top) NASA – (bottom) Roger Ressmeyer Starlight, Science Photo Library; page 31 – Martin Marietta; page 32 – NASA, Science Photo Library; page 33 – (top) British Aerospace Space Systems Ltd – (bottom) Deutsche Aerospace; page 34 – Fotokhronika Tass; page 35 – Alan Chinchar, NASA; page 36 – (top) NASA, Science Photo Library – (bottom) European Space Agency/NASA; page 37 – Genesis Space Photo Library; page 38 – NASA, Science Photo Library; page 39 – (top left) Fotokhronika Tass – (top right and bottom) NASA, Science Photo Library; page 40 – David A. Hardy; page 42 – (left) A. de Menil, Science Photo Library – (right) Malcolm Fielding/The BOC Group PLC, Science Photo Library; page 43 –Fotokhronika Tass – (inset) Black and Decker.

Contents

Int
Space sci
when pe
wonder
sands o
as we k
transfo
cise th
the sp
   Un
could
meas
fortu
or o
tha
un
to
tw
p
d
g

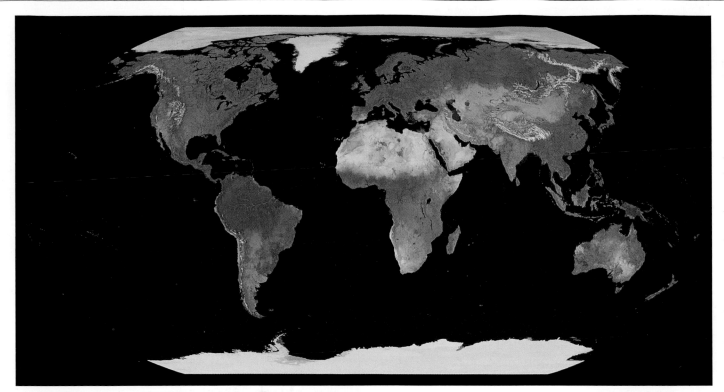

This striking image of the Earth was obtained by piecing together thousands of separate pictures taken by TIROS satellites. Only cloudless images were included, to give a clear view of the ground and oceans over the whole planet.

and gold has been discovered in Australia. The oceans are also closely monitored by satellite because they may affect our climate.

Satellites used to explore Earth in this way are called Earth Resources Satellites.

Science in space

There are many other types of satellites, each designed and built to do a particular job. They include communications, navigation, weather, and astronomy satellites. Less is known about the military satellites that many countries use to spy on other countries. Another type of spacecraft, the deep space probe, is designed to travel far beyond the space around Earth to visit other planets in our solar system.

Gravity, the force that pulls us down onto Earth's surface, can cause undesirable effects in some scientific experiments and manufacturing processes performed on Earth. One solution is to eliminate the effects of gravity by carrying out this type of work in permanently orbiting spacecraft called space stations. These large structures and their crews need regular supplies brought from Earth by space shuttles. NASA, the U.S. space agency, has developed a shuttle as the first step in building and operating a space station.

All of these different types of spacecraft and the ways in which they will be used in the future are discussed in this book. You may like to consider your own views on the questions that are featured in the **Space for thought** boxes. Words in bold type are defined at the end of each section.

atmosphere – the thin layer of gases that surrounds Earth, composed mainly of nitrogen and oxygen.

radiation – rays of energy and particles reaching Earth from outer space.

galaxies – collections of millions upon millions of stars traveling through space together. Our solar system is part of one galaxy called the Milky Way. Galaxies are separated by unimaginably enormous distances.

payload – the equipment carried by a rocket or satellite.

sensors – the part of a satellite's payload that reacts to whatever it is designed to detect — heat, light, or radio energy, for example. This reaction is converted to an electrical signal and radioed back to Earth.

Eye in the Sky

The satellite, the basic tool of the space scientist, is merely a mobile platform to which scientific instruments are attached. It requires a powerful rocket engine to reach its planned **orbit** and smaller engines to move within that orbit. It also needs some means of producing electrical power for on-board equipment. Most satellites use panels covered with solar cells. The cells convert sunlight directly into electricity. When the satellite passes into Earth's shadow, the solar panels cease to generate electric current and batteries take over. When the satellite can use solar power again, the batteries are recharged by the solar cells, ready for the next night period.

The anatomy of a satellite

The European Earth Resources Satellite, ERS-1, is the forerunner of a new generation of spacecraft that will study Earth, its atmosphere, and the space around Earth. ERS-1, launched in 1991, carries **radar** instruments that are able to take measurements of the land, sea, and ice below, whether in daylight or darkness.

Rocket engines

Rocket engines work by burning fuel, often liquid hydrogen, in a **combustion** chamber. The hot gases that are produced expand rapidly and rush out of the rocket at great speed through a nozzle. The force of the hot gases traveling in one direction produces the thrust that pushes the rocket in the opposite direction. To make liquid hydrogen burn, it must be supplied with oxygen. This is stored in the rocket as an oxidizer.

Booster rockets, used to increase the power of a liquid-fueled rocket so that it can lift heavier payloads, often use a solid fuel like a large firework. Tiny rocket engines called thrusters, used to make small adjustments to a spacecraft's position, often use fuels that burn when they mix. They do not need any spark or flame to ignite them.

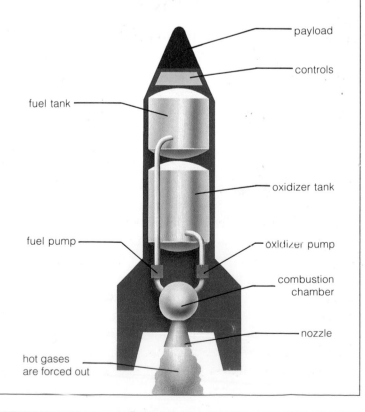

payload

controls

fuel tank

oxidizer tank

fuel pump

oxidizer pump

combustion chamber

nozzle

hot gases are forced out

Ariane 4 (left). The launch of the European Space Agency's *Ariane* rocket from Kourou in French Guiana, South America.

Launching a satellite

Most satellites are launched by expendable (use once and throw away) rockets. The rocket is made from three or more separate sections, or stages, each with its own fuel tanks and rocket engines. The first stage is fired to lift the rocket and its payload off the ground. When its fuel is used up, it is detached from the rest of the rocket and falls back to Earth. This reduces the weight of the rocket and enables the next stage to push the rocket farther and faster. When it is traveling at the correct speed, height, and direction, the casing around the satellite opens, and the satellite emerges into space.

Orbiting Earth

A satellite may be placed in one of three main types of orbits — polar, low Earth orbit around the equator, or geosynchronous. A satellite in a polar orbit travels from pole to pole while Earth spins beneath it. This enables the satellite to look at almost any part of Earth. Earth Resources Satellites and military spy satellites are often placed in polar orbits for this reason.

Equatorial regions are best served by low-altitude orbits around the equator. Other satellites are put into orbit at higher altitudes. The higher a satellite orbits around the equator, the longer it takes to fly from horizon to horizon. At a height of 22,000 miles, the satellite orbits at the same speed as Earth rotates. It appears to hang over the same spot on Earth. This is ideal for communications satellites and some weather satellites because they are constantly in contact with the same part of Earth and they do not need to be tracked by moving **antennae.** This is known as a geosynchronous, or geostationary, orbit.

Geostationary orbits are not the best for communications near Earth's poles: from either pole a satellite flying over the equator appears low down at the horizon, and Earth's atmosphere can distort the signal between satellite and receiver. The former Soviet Union used a different orbit for its communications satellites. Its Molniya satellites were placed in an elliptical orbit that takes them high over its territory.

SATELLITE ORBITS AROUND EARTH

Satellites are placed in three main types of orbit around Earth. 1. Polar orbit around Earth's poles. 2. Low Earth orbit around the equator. 3. Geosynchronous or geostationary orbit at an altitude of 22,000 miles. The Soviet Molniya satellites were placed in an elliptical orbit that took them high over the former Soviet Union.

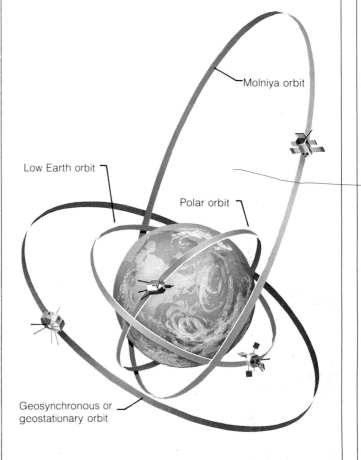

Molniya orbit

Low Earth orbit

Polar orbit

Geosynchronous or geostationary orbit

orbit – the path followed by an object such as a spacecraft or a moon as it circles a planet, or the path of a planet around a star.

radar – (Radio Detection and Ranging) a system used to locate objects and draw maps of a planet's surface by sending out bursts of radio waves and analyzing any reflections that bounce back.

combustion – the point at which substances react with oxygen to produce a rise in temperature and start burning.

equatorial – regions near the equator, an imaginary line around Earth's middle, midway between the poles.

antenna – a metal wire, frame, plate, or dish used to detect radio signals.

Remote Sensing

We normally observe our surroundings by using our five senses — sight, hearing, touch, smell, and taste. Our eyes can detect only a tiny fraction of the sun's radiation, which we call light. If we could see other **wavelengths**, such as radio waves, infrared, and ultraviolet, the world would look very different because this invisible radiation carries extra information about the world around us. The use of sensors, artificial "eyes" that are sensitive to invisible radiation, to learn more about distant objects is called remote sensing. It is an increasingly important branch of space science.

To most people, the most familiar use of remote sensing is weather forecasting. But remote sensing has many other uses. Satellites can detect effects that are difficult to study on the ground. For example, the slight changes in the color of crops and trees that indicate the onset of a disease can often be picked up by satellites (see page 12). Satellites are also used to study other problems, such as the greenhouse effect (see page 13) and damage to the ozone layer (see page 14).

Weather forecasting

Until the 1960s, meteorologists (weather scientists) had to piece together the structure of weather systems from observations made by isolated weather stations on Earth. It was like trying to do a huge jigsaw puzzle with half the pieces missing. Since 1960, when the first weather satellite was launched (TIROS-1), forecasters have been able to see pictures of weather systems developing across the globe.

Weather satellites generally occupy one of two types of orbit. The United States uses TIROS (Television and Infrared Observation Satellite), which travels in a polar orbit. TIROS can photograph almost any point on Earth twice a day, making it possible to piece together a picture of the world's weather every 12 hours. European Meteosat, American GOES (Geostationary Operational Environmental Satellites) and Japanese GMS (Geostationary Meteorological Satellites) occupy geostationary orbits, watching the weather over one area.

Modern weather satellites send down visible light pictures of Earth with clouds moving

Satellite views. This severe storm (left) was tracked across the Atlantic Ocean by the NOAA 9 satellite in 1987. More unfamiliar satellite views show the amount of water vapor in the atmosphere (below) and oil in the Arabian Gulf in 1991 (right). The oil slick is highlighted in red. The oil threatened many wildlife species, including the green turtle (inset).

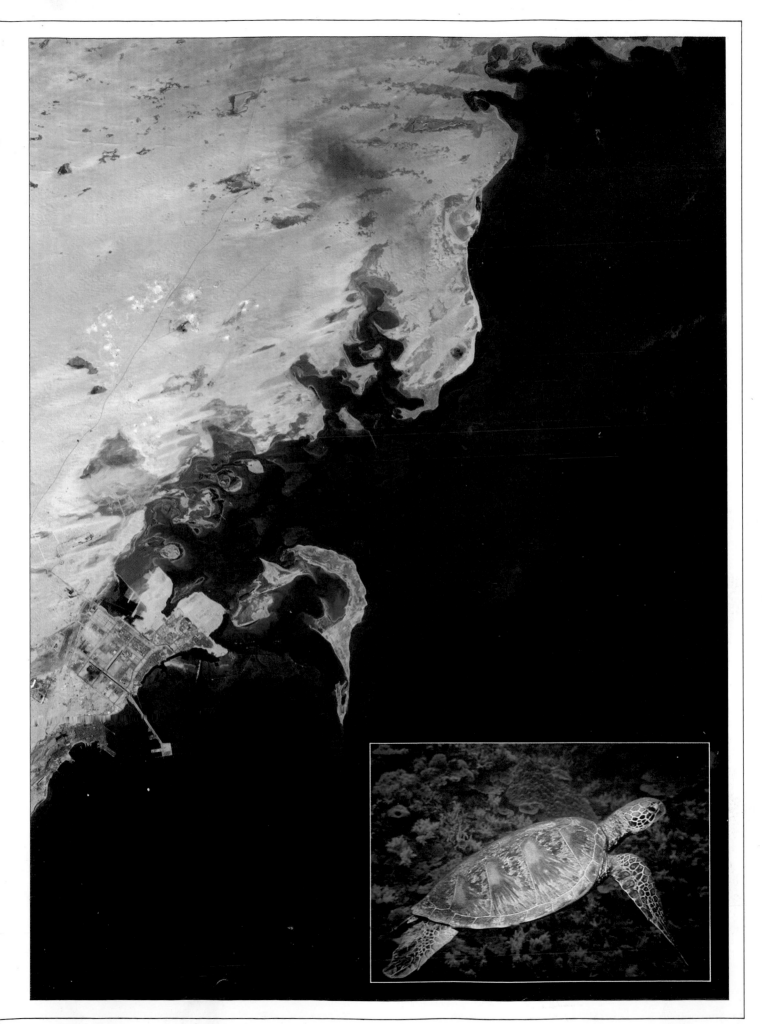

across it. They also obtain other very useful information. For example, detectors sensitive to infrared radiation can measure the temperature of clouds and how much water they contain, useful for predicting rainfall. Land and sea temperatures can also be measured.

In the United States the next series of GOES, called GOES NEXT, will improve on its predecessors by being able to view smaller areas with infrared sensors. The best pictures available from GOES are of areas five miles in diameter. GOES NEXT will produce detailed images of areas only 2.5 miles across. GOES NEXT will also take less time to build up its images line-by-line, producing one picture per minute — 15 times faster than the present environmental satellite — GOES.

Satellite receiving equipment has been miniaturized so much that it is now possible to make a desktop weather station. An antenna and a receiver tune in to the satellite's radio signal, and a desktop computer turns the coded signal into a weather picture on the computer screen. These systems are particularly useful for providing farmers and construction companies with updated forecasts — weather forecasts that cover the next few hours.

Natural disasters

Storms, erupting volcanoes, droughts, and floods are all natural disasters that can cause widespread suffering and death, disrupting industrial and commercial life, and leaving in their wake a trail of destruction that is expensive and difficult to clear up. They are difficult events to predict and warn people about because they tend to happen suddenly and apparently without warning. Satellite pictures can provide some warning of these disasters, and assist the rescue operations, too.

Tracking a Storm

Weather satellite pictures show storms and hurricanes approaching land and indicate how strong they are, giving time to warn coastal areas. Tropical storms over the Atlantic Ocean, for example, are tracked by satellite as they gain strength. As the storm or hurricane approaches land, the part of the coast it will strike is predicted and, if necessary, the residents are evacuated. Unfortunately, some of the most destructive storms occur in parts of the world where communications are poor, such as Bangladesh, and people often cannot be warned in time.

Satellite observations of how water is used over wide areas can help governments to cope with drought (left), and to improve water supplies through irrigation schemes. This satellite picture of the Nile River (below) shows the irrigated land in red. Agriculture in the Nile delta relies on irrigation to grow crops such as cotton, corn, and clover.

Hawaii 1986. By monitoring a volcano by satellite, volcanic eruptions can sometimes be predicted.

Drought

Drought can also be dangerous as a killer of vegetation, crops, animals, and people. Using satellite-mounted radar to monitor the shape of the land surface, rock and soil types, and patterns of vegetation and land use, produces valuable information about how water behaves on and under the land. This enables scientists to warn of approaching drought, and to give planners the information they need to dig new wells, build new reservoirs, and change the ways in which land is used in order to minimize the effects of drought.

Volcanoes

Volcanoes undergo a series of changes before they erupt. As molten rock builds up underneath a volcano, the sides of the mountain swell up. French scientists have detected this swelling in satellite studies of Europe's most active volcano, Mount Etna in Sicily. Radio signals transmitted by American Global Positioning System (GPS) satellites (see page 19) were picked up by receivers on the side of the mountain and used to calculate their positions. If the ground moved even slightly, the receivers moved with it and the movement was detected by the satellite system. Heat sensors on board Landsat satellites have also been used to monitor temperature changes in active volcanoes in Chile and Ethiopia. Sudden changes in the heat output of a volcano may indicate that the volcano will soon erupt.

Most earthquake monitoring is carried out on the ground, but the techniques used to detect movements on Mount Etna could also be used to detect the tremors that shake the ground before an earthquake.

Accurate prediction on natural disasters to within a few hours or even days is still very difficult. But the mass of information produced by satellite monitoring may hold the vital clues that will enable scientist to improve the accuracy of these predictions in the future.

Destroying the destroyers

Insects cause billions of dollars worth of damage to forests and crops and serious diseases in farm animals and humans. Insects themselves are too small to be seen from space, but their habitats and the damage they cause can be detected by satellite.

Satellites have helped to combat a tick that causes animals diseases on the Caribbean island of St. Lucia. The tropical bont tick lives where the mesquite bush grows and these areas can be spotted by satellite. Knowing this, the local authorities have devised a way of destroying the tick's habitat and the tick itself.

In South America, a pest called the cotton boll weevil prevents cotton plants from developing cotton. During the 1980s, the weevil spread south from Central America through Brazil toward Paraguay. The American Land-

sat and French Spot satellites were used to map cotton-growing areas in Paraguay so that strips of land in the path of the weevil could be cleared to slow or halt its progress.

In North America, snails living in marshes in Louisiana carry a parasite which, when eaten by cattle, develops into a creature called a liver fluke. Satellite images have helped to highlight areas where the marshy habitats of the snails meet the sandy land where the cattle live. This is where the parasite is passed from snails to cattle.

Without satellites, finding places where pests breed and feed is time-consuming and costly. Satellite images can provide enough information to guide pest control teams. Satellites can also be used to alert farmers to areas of stress, where crops are suffering because of lack of water, or from disease.

Satellite pictures are used to compare healthy crops (bottom) and stressed crops (top) which may be suffering from drought or disease.

Crops such as the wheat shown here (left) suffer considerable damage from insect pests and diseases. The cotton boll weevil (above) attacks cotton crops.

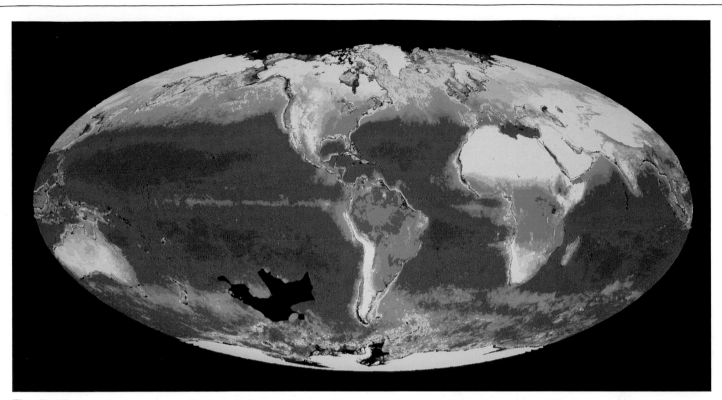

The distribution of vegetation on land and phytoplankton in the oceans is shown in this false-color satellite image. The densest phytoplankton is coded red, decreasing through yellow and blue to pink, the least dense.

The greenhouse effect

Storms, volcanoes, floods, and plagues of insects are natural disasters, but some problems on Earth are caused or worsened by human activities. Many scientists are concerned about two problems in particular — the greenhouse effect and the thinning of a layer of Earth's atmosphere called the ozone layer. Satellites are helping to monitor both of these problems.

Earth soaks up energy from the sun. This energy is trapped by Earth's atmosphere for a time, warming the planet before it escapes back to space. Some gases, such as carbon dioxide, are particularly good at trapping heat. If the amount of carbon dioxide in the atmosphere increases, the sun's heat takes longer to escape from the atmosphere, and Earth's surface temperature rises. This is called the greenhouse effect. It could have a disastrous effect on our climate, turning agricultural land into desert and flooding coastal areas. Many scientists believe human activities are upsetting the balance of atmospheric gases that control our climate.

When living things die and rot, or if they are burned, carbon escapes from their dead cells into the atmosphere as carbon dioxide gas. Burning fossil fuels, such as coal and oil, also releases carbon into the atmosphere in the form of carbon dioxide. Plants absorb carbon

dioxide, but if there are fewer plants because of forest clearance, more carbon dioxide stays in the atmosphere. Forest clearance is also monitored by satellite.

It is well known that the oceans that cover most of our planet affect our weather in many ways. The oceans are also now known to be the world's largest absorbers of carbon dioxide. Understanding how gases are exchanged between the sea and the air is therefore very important in predicting the progress of the greenhouse effect and our future climate.

Satellite images have revealed the importance of microscopic plant-like organisms called phytoplankton. They take in carbon dioxide dissolved in the water from the atmosphere and turn it into carbon in their own cells. Satellite images have shown dramatic changes in the numbers of phytoplankton near the ocean surface from season to season, and this affects the amount of carbon dioxide absorbed from the atmosphere. More satellites with ocean sensors are due to be launched in the late 1990s.

Not all scientists believe that the greenhouse effect will result in a warmer climate. Some think that if the land and seas begin to warm, more water may **evaporate** from them, creating clouds that would reflect more of the sun's energy back to space and cool Earth again.

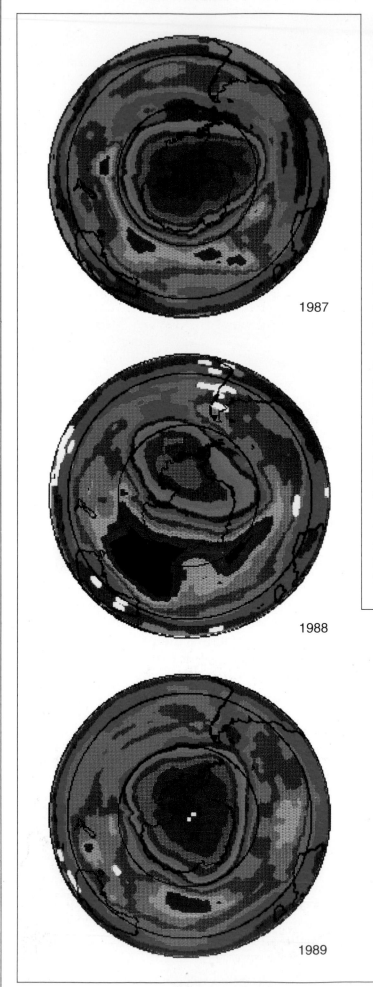

1987

1988

Protecting the ozone layer

Since the early 1980s, scientists have known that an important layer of the atmosphere, the ozone layer, was becoming very thin in places, especially at the poles. Ozone is a type of oxygen whose molecules contain three oxygen atoms instead of the normal two. Made by the action of intense sunlight on oxygen high in the atmosphere, ozone is very important because it absorbs ultraviolet radiation and prevents most of it from reaching Earth's surface. The small amount that does reach us is responsible for producing a suntan. More intense ultraviolet radiation can cause illnesses including skin cancer and the clouding of the eye's lens, called a cataract. It could also kill plankton in the sea and reduce crop yields.

Damage to the ozone layer is partly the result of human actions. In the mid-1970s, scientists suggested that gases called chlorofluorocarbons, better known as CFCs, used in aerosol spray cans and refrigerators could float up to the ozone layer. There, sunlight broke the CFCs down, and the chlorine gas released destroyed the ozone around it. Because of this, many countries have banned CFCs in aerosol cans.

The ozone layer. These satellite pictures show holes in the ozone layer over the South Pole recorded on October 3 between 1987 and 1990. The pink and deep pink colors indicate extremely low ozone levels.

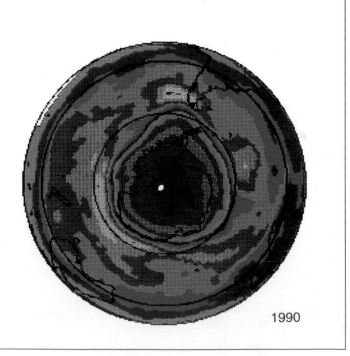

1989

1990

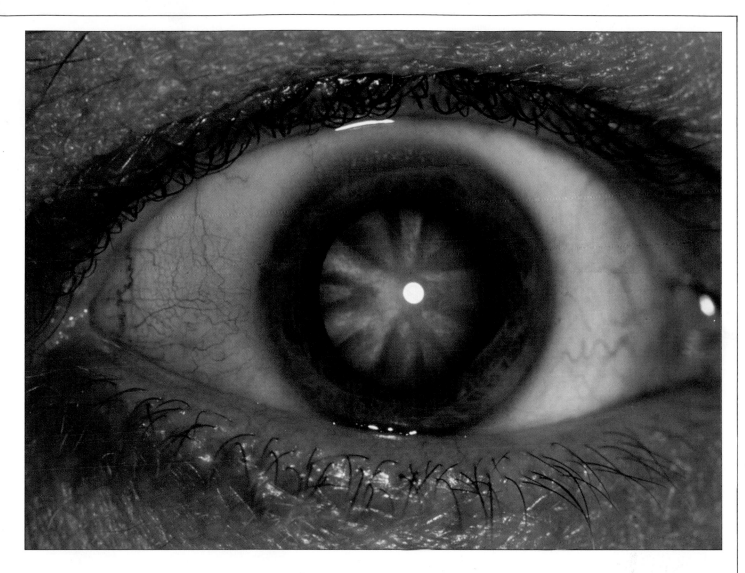

The ozone layer has been monitored since 1978 by an instrument called the Total Ozone Mapping Spectrometer (TOMS) on board the satellite *Nimbus 7*. In September 1991, a far more sophisticated satellite, the Upper Atmospheric Research Satellite (UARS) was put into orbit. It will provide scientists with detailed information about the nature and chemistry of the ozone layer.

Earthwatch

At a conference in 1989, leaders of the seven richest nations agreed to begin the most thorough study of Earth yet undertaken. The outcome was the Earth Observing System (EOS). Information collected by a fleet of new satellites will help scientists to understand our weather, volcanic activity, polar snow and ice behavior, and many other aspects of our planet. This may help scientists to predict what will happen to our planet in the future. The first EOS satellites are expected to be launched in 1996.

A cataract in the eye can be caused by, among other things, increased exposure to ultraviolet radiation. In this picture, the wheel spoke appearance is typical of a developing cataract which will gradually cloud the sufferer's vision.

Space for thought
● If attempts to protect the ozone layer fail, and more of the sun's radiation reaches Earth, how might people have to change their way of life to stay healthy?

wavelengths – radio waves, infrared and ultraviolet rays are all part of the electromagnetic spectrum. They consist of electrical and magnetic vibrations. The distance between two neighboring vibrations is called the wavelength.
evaporate – the change of state from a liquid to a vapor.

Communications

Satellites have revolutionized communications. When the world's first commercial communications satellite, *Early Bird* (also known as *Intelsat 1*), was launched in 1965, it could relay one black and white television channel or 240 telephone conversations across the Atlantic Ocean. The latest in the Intelsat series of satellites, *Intelsat 6*, can relay up to 120,000 telephone calls and three color television channels. The satellite *Olympus 1*, launched in 1989, is a new type of high-power communications satellite that will be able to relay up to 40 color television channels or a quarter of a million telephone conversations.

In the future, communications satellites are likely to become even bigger and more powerful. The more powerful a satellite, the smaller the equipment needed on the ground to communicate with it. This will make satellite communications, or satcoms, available to many more people.

Keeping in touch

Satellite communication is already available to telephone users through the international telephone network. Using a car telephone, portable computers, and **facsimile transmission (fax)** machines, a passenger traveling in a car can now send voice messages, printed text, and computer data by satellite to an office or even another car almost anywhere in the world today.

Not everyone is able to make conventional telephone calls. In some circumstances special direct-to-satellite services are the only means of communication. In the aftermath of an earthquake, rescue workers trying to locate injured people buried in the rubble may use portable satellite equipment if the telephone system in the affected area is not working. Portable satellite terminals also enable television news crews to send back the sounds and images of events in other parts of the world as they are happening.

In the future, these special satellite services will be available to more people. Truck drivers, for example, can already fit satcoms sets to their trucks, enabling them to keep in touch with their base wherever they are in the world. And improvements in radio communications between airplanes and the ground via satellite mean that passengers can now make telephone calls while in flight.

Spy satellites

Military forces in the developed world have become increasingly dependent on satellites for information-gathering, communications, weather forecasting, and navigation. Information-gathering satellites, better known as spy satellites, can use their own rocket engines to change orbit so that they can turn their cameras and radio receivers on any part of Earth's surface. The United States started launching a

series of spy satellites called Keyhole in 1962. The current Keyhole satellites, KH-11 and KH-12, send very detailed video images of their targets to Earth by radio. As a result of its low orbiting height, a Keyhole satellite has a maximum life of only two years. **Friction** between the satellite and the thinnest outer fringes of Earth's atmosphere gradually slows the satel-lite. As it loses speed, it also loses height. It is repeatedly nudged up to a safe height until its rocket engine runs out of fuel, then it eventually spirals down into Earth's atmosphere where it burns up.

In the future, because of new techniques demonstrated by the American space shuttle, there may be no limit to the length of time a

Office in a car. This car has been fitted with a mobile telephone linked by radio to the international telephone network. It also has a portable computer, a printer, and a fax machine.

Rescuing the *Westar 6* communications satellite. Astronaut Joseph P. Allen is standing on the mobile foot restraint of the shuttle *Discovery.* Satellites can be repaired in space, or returned to Earth.

Intelsat 6 **(left).** One of the world's most advanced communications satellites, *Intelsat 6* stands as high as a four-story building and weighs three tons.

spy satellite can stay in orbit. In 1984, shuttle astronauts repaired a satellite in orbit, practiced refueling a satellite, and rescued two satellites that had been placed in the wrong orbits and returned them to Earth. Using these operations, military satellites could be refueled and repaired in space. It is thought that KH-12 satellites were designed to be refueled in space.

Military communications

Good communications between commanders and their battle groups, planes, ships, and submarines enable military forces to be guided directly to where they are needed. Communications equipment is now so small and easy to carry that individual soldiers can set up a small dish antenna and use a portable satellite telephone to communicate with one another.

Submarine communications pose particularly difficult problems. The role of a nuclear submarine is to hide unseen in the ocean. Yet, because radio signals do not penetrate water well, the submarine must rise

The Global Positioning System (GPS) is used by military and civilian shipping. Yachts and fishing boats are fitted with a Navstar GPS receiver (inset), which tracks up to eight Navstar satellites and gives the boat's position to within 50 feet.

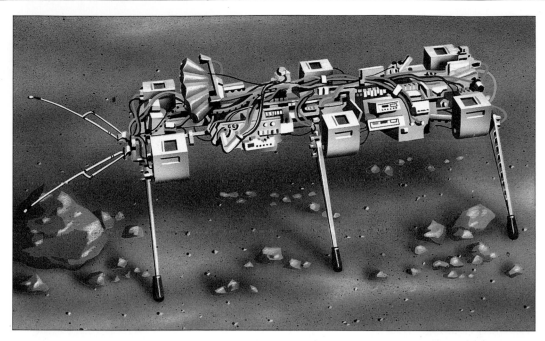

close to the surface, where it can be easily detected, in order to communicate. Extremely Low Frequency (ELF) radio waves penetrate water, but ELF transmitting stations on land are enormous and present an easy target to air attack. A beam of blue-green light from a **laser** penetrates water better than radio waves and also carries more information more quickly than ELF radio. In the future, laser satellites may be used to improve submarine communications.

Navigation

The **navigation** satellite is becoming one of the most important types of military satellite. It works by means of a principle called the Doppler effect. If a satellite travels towards a receiver on Earth, the radio signal received from it increases in **frequency** in the same way as the whine of an approaching aircraft rises. If the positions, directions, and speeds of several satellites are known, the Doppler effects of their signals enable the position of the receiver to be calculated very accurately. The United States has launched a series of Navstar satellites to form a Global Positioning System (GPS). It enables anyone with a small GPS receiver to work out where they are to within 50 feet. It is used mainly by military and civilian sailors and pilots, and by soldiers.

Of course, as satellites become vital to military campaigns, it also becomes more important to be able to destroy enemy satellites. A satellite's optical sensors can be blinded by a laser. The satellite itself can be destroyed by an anti-satellite missile fired from an aircraft, or by an anti-satellite satellite. This moves close to an enemy satellite and explodes. The powerful burst of radio energy from a nuclear explosion can also burn out delicate electronic circuitry. Military satellites must do their demanding work and defend themselves against these threats.

Reducing the cost of spaceflight

Space programs are extremely expensive, often costing billions of dollars. Only the wealthiest governments can fund a space program, resulting in relatively few opportunities for scientists to get their experiments into space. If spaceflight were cheaper, more scientists could carry out more experiments in space. Changing from throwaway rockets and spacecraft to reusable equipment is one way of cutting the cost of each launch, but there is also another way.

Smallsats

Since the first satellites were launched in the 1950s, the development of larger and more powerful rockets has enabled space scientists to place larger and larger satellites in space. *Sputnik 1* weighed only 184 pounds. The largest satellites in orbit now weigh up to 15 tons. This trend toward ever-bigger satellites is now being questioned.

Advances in the miniaturization of electronics and mechanical parts mean that scientists are now able to build spacecraft that are smaller, lighter, cheaper, and also capable

MICROSATS

This design for a microsat measures only 6 inches across. It carries two tiny video cameras, one to record wide-angle views and the other for close-ups. Scientists predict that the next generation of microsats will be even smaller — about the size of a soda can.

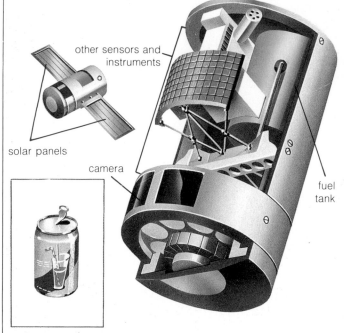

other sensors and instruments

solar panels

camera

fuel tank

Launching smallsats

Smallsats and microsats are cheaper to launch because smaller rockets can be used. But these spacecraft are so small that launch systems other than rockets are also possible. One launch system being studied now would fire smallsats into orbit by using a large gun, called a coil gun. This gun is made up of a series of powerful **electromagnets** surrounding a tube. The satellite is inserted into the tube. The electromagnets are turned on, one after the other, pulling the satellite along the tube and accelerating it to a speed of between 9,000 and 12,000 miles per hour. If the coil tube is aimed carefully, this is fast enough to place the satellite on a path that will take it into orbit around the Earth. Scientists hope to build the first coil gun within the next ten years. Until then, smallsats will be launched from Earth by rockets, or released into orbit in space from the payload bay of the space shuttle.

of doing more than ever before. Known as smallsats (also lightsats or cheapsats), they will weigh tens of pounds instead of tons, and they will cost less than two million dollars each to produce. Smallsats will not replace all large satellites. For example, communications are probably best served by a few very large satellites, and so large comsats will continue to be built. Other scientific satellites must be large in order to accommodate the large amount of equipment that they have to carry.

Microsats

The first smallsats are being produced now in Europe and the United States. Scientists are already predicting that the second generation of smallsats will weigh only 2 pounds, or even less, and be similar in size to a soda can. These microsats would be used differently from a single huge satellite. A swarm of dozens of microsats launched at the same time would make observations from lots of different places.

If one large satellite is faulty or destroyed in an accident, the whole mission and all the experimental work that the satellite would have done is lost. However, if one microsat in a group of five, ten, or even a hundred is lost, the mission can still proceed successfully.

Space for thought

● If your school could afford to send a microsat into space, what experiment would you like it to carry, and why?

facsimile transmission (fax) – a system for sending printed text, drawings, handwriting, and pictures by telephone through a fax machine.

friction – a force produced when things rub against each other. Friction between a rocket and air rushing past it slows the rocket down.

laser – a device that produces an intense beam of light.

navigation – finding a route from one place to another, and being able to plot your position upon the earth's surface.

frequency – the number of cycles, or waves, that pass any point in a second. The waves could be water waves, sound, light, or radio waves.

electromagnets – magnets made from a coil of wire wound around an iron bar. When an electric current flows through the coil, the bar becomes a magnet.

THE COIL GUN
The coil gun would use the power of magnets to fling a satellite out through its barrel fast enough to take it into orbit around Earth.

Exploring Space

People have always wanted to explore their surroundings. Until the space age began, there was no way to explore the solar system apart from looking at it through telescopes. Now, spacecraft can be sent out to the planets. Some of the most spectacular photographs taken this century show the surface of Mars taken by Viking probes, and close-up views of the planet Jupiter and some of its many moons taken by Voyager probes. Radar sensors carried by the probe *Magellan* have penetrated the thick atmosphere of Venus to picture its surface. But planetary probes do much more than take pretty pictures with their on-board cameras. They also carry instruments that listen to naturally-produced radio signals coming from the planets (from thunderstorms, for example) and measure the planet's magnetic field. The photographs and measurements enable scientists to work out the structure of the planet, how hot or cold it is, what chemicals are found in its atmosphere, and much more.

Magellan. The deep space probe is checked by its manufacturers (above). On May 4, 1989 it was released from the payload bay of the space shuttle *Atlantis* toward Venus (below).

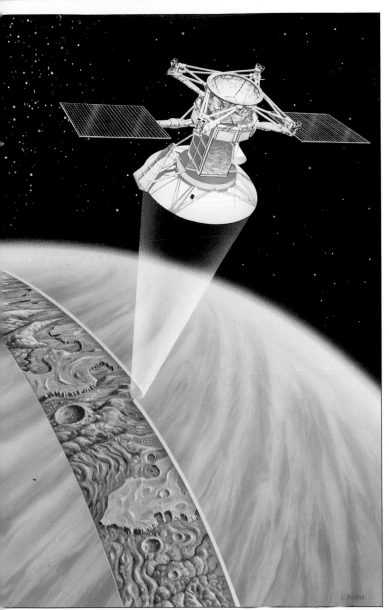

Deep space probes

The scientists and engineers who design and build deep space probes have to solve several problems. In the outer reaches of the solar system there is not enough sunlight to generate electricity with solar cells. By the time a probe reaches the orbit of Jupiter, for example, the sun's energy is only one twenty-fifth of its intensity at Earth orbit.

Deep space probes traveling beyond the orbit of Mars use a different system to generate power. They carry nuclear power supplies, called radioisotope thermoelectric generators (RTGs). They contain plutonium, which is **radioactive** and breaks down naturally, releasing radiation that is converted into electricity.

All spacecraft receive commands from Earth and send data back to Earth by radio. As a satellite travels farther away from Earth, the radio signals received by the spacecraft and by its ground control center become weaker. Deep space probes are, therefore, usually fitted with large dish antennae to help concentrate the transmitted and received signals. The *Ulysses* spacecraft that will investigate the sun in the 1990s has a 5-foot dish. The signals from its 20-watt transmitter (less than some light bulbs) will be picked up from 600 million miles away by the large dish antennae of **NASA**'s deep space network.

Several missions to the planets are planned for the next ten years. In 1996, the *Cassini/*

Safely in orbit around Venus (above) with its solar panels unfolded, *Magellan's* radar sensor picks up images through the planet's thick atmosphere and starts mapping the surface. The thin strip on the left of the picture (right) is the first image taken by *Magellan* of the surface of Venus. The small white box on the larger picture shows the best view of the same area that can be obtained from Earth.

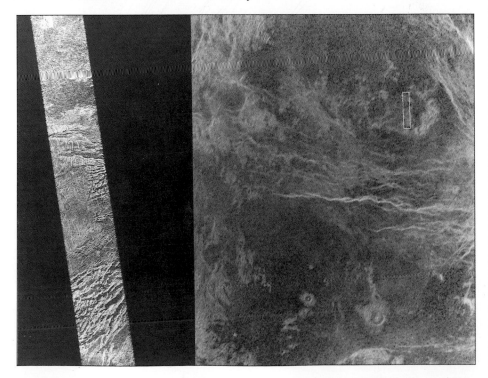

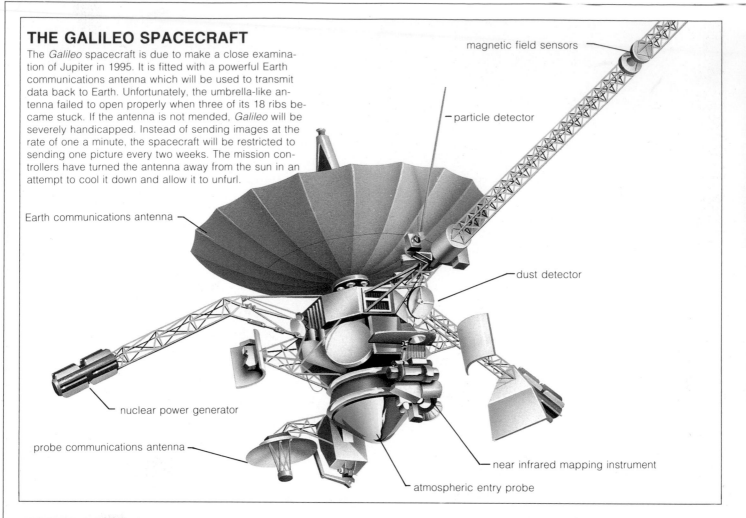

THE GALILEO SPACECRAFT

The *Galileo* spacecraft is due to make a close examination of Jupiter in 1995. It is fitted with a powerful Earth communications antenna which will be used to transmit data back to Earth. Unfortunately, the umbrella-like antenna failed to open properly when three of its 18 ribs became stuck. If the antenna is not mended, *Galileo* will be severely handicapped. Instead of sending images at the rate of one a minute, the spacecraft will be restricted to sending one picture every two weeks. The mission controllers have turned the antenna away from the sun in an attempt to cool it down and allow it to unfurl.

magnetic field sensors

particle detector

Earth communications antenna

dust detector

nuclear power generator

probe communications antenna

near infrared mapping instrument

atmospheric entry probe

The *Galileo* probe will become the first Earth-made object to enter the atmosphere of an outer planet when it reaches Jupiter in 1995. At left, the probe's heat shield is shown falling away after entry into Jupiter's atmosphere.

Huygens mission will send a spacecraft to orbit Saturn where a probe will separate from the spacecraft and fly to Saturn's largest moon, Titan. There, it will land and send back pictures from the surface. The *Galileo* spacecraft, already on its way to Jupiter, will drop a probe into Jupiter's atmosphere. On its way to Jupiter, *Galileo* sent back the first close-up picture of the **asteroid** Gaspra.

The sun and stars

Our nearest star, the sun, will come under increasingly close attention in the 1990s and beyond. The sun is vital to life on Earth. Without the sun's gravitational force, there would be no solar system. Without exactly the right amount of warmth and sunlight, there would be no life on Earth. The sun appears to shine unchanging day after day, but it does change. Scientists now believe that very small variations in the sun can have major effects on Earth.

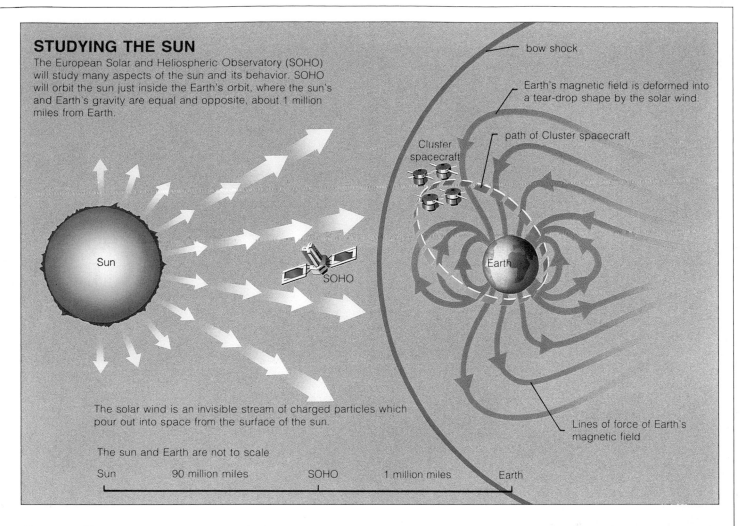

STUDYING THE SUN

The European Solar and Heliospheric Observatory (SOHO) will study many aspects of the sun and its behavior. SOHO will orbit the sun just inside the Earth's orbit, where the sun's and Earth's gravity are equal and opposite, about 1 million miles from Earth.

Sun

SOHO

Cluster spacecraft

Earth

bow shock

Earth's magnetic field is deformed into a tear-drop shape by the solar wind.

path of Cluster spacecraft

Lines of force of Earth's magnetic field

The solar wind is an invisible stream of charged particles which pour out into space from the surface of the sun.

The sun and Earth are not to scale

| Sun | 90 million miles | SOHO | 1 million miles | Earth |

Sun studies

The European, American and Japanese space agencies have joined forces to send a fleet of spacecraft to study the sun. One of them, the European Solar and Heliospheric Observatory (SOHO), will study the sun's structure and the particles that stream away from the sun in all directions, forming a solar wind. SOHO will orbit the sun just inside Earth's orbit, at a point where the gravitational forces of Earth and the sun cancel each other out. SOHO's instruments will observe how the sun vibrates. In the same way as earthquakes give scientists clues about Earth's internal structure, "sunquakes" should provide important clues about the makeup of the sun.

The European Solar and Heliospheric Observatory will also study sunspots, dark marks on the sun's surface. The spots look dark only because they are about 2700° cooler than the surrounding 11,000°F surface material. Each spot is the center of an intense **magnetic field,** 1000 times stronger than the sun's general magnetic field, that has burst through to the sun's atmosphere from its interior.

The number of sunspots changes according to a regular 11-year cycle. Scientists believe that Earth's climate may be linked with this sunspot cycle. The areas around sunspots are very active. Explosions called prominences and flares can send jets of particles out from

An ultraviolet image showing a giant solar prominence erupting near a sunspot.

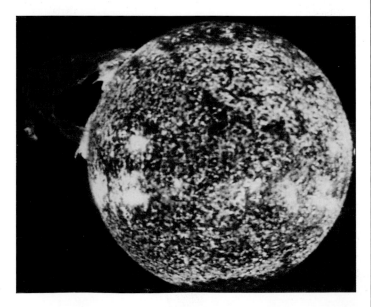

25

The aurora borealis, or Northern Lights, photographed in Alaska.

the sun at great speed. Some are trapped by Earth's magnetic field, causing the colorful light displays in the sky near the poles called the aurora borealis in the north, and the aurora australis in the south.

While SOHO studies the sun itself, several other spacecraft, called the Cluster mission, will study the effect of the sun on Earth's magnetic field. Four spacecraft will take measurements of the regions between Earth and the sun to find out what happens to the solar wind when it meets Earth's magnetic field.

Another joint European Space Agency (ESA) and NASA spacecraft, called *Ulysses*, will become the first spacecraft to orbit the sun's poles. *Ulysses* was launched on October 6, 1990, but it will not pass over the sun's poles for the first time until May 1994. On its way, it will become the fastest human-made object in the universe, traveling at 7 miles per second. *Ulysses* carries nine instruments that will study the sun's magnetic field, the solar wind, and

the bursts of radio energy that the sun produces from time to time.

The Hubble Space Telescope

In the near future, some of the most exciting new information about distant stars and galaxies is expected to be provided by a giant orbiting telescope, the Hubble Space Telescope.

Named after the American astronomer, Edwin Hubble (1889 – 1953), the telescope is over 42 feet long by 13 feet across, and weighs 12 tons. Its circular main mirror, used to collect light from distant stars, is 8 feet across. Several telescopes on Earth have much larger mirrors. However, the Hubble telescope, above Earth's atmosphere, observes stars and galaxies 50 times fainter than those visible using Earth-bound telescopes. Those telescopes are hindered by Earth's atmosphere.

The distances between stars and galaxies are so great that they are measured in light-years. A light-year is the distance that light travels in

a year. At a speed of 180,000 miles per second, this is equivalent to almost six trillion miles per year. As the light from such distant objects takes millions upon millions of years to reach Earth, looking far into space means looking back in time. The Hubble Space Telescope should give astronomers more clues about conditions in the early universe and even how the universe began.

Unfortunately, soon after its launch from the space shuttle in April 1990, a fault was discovered in the telescope. It could not be focused sharply because its main mirror was the wrong shape. The error is less than 1 ten-thousandth of an inch, but it is enough to prevent the telescope from producing clear images.

The main mirror cannot be replaced, but since the telescope was designed to be serviced and repaired in space, astronauts will replace one of its instruments with a new assembly of mirrors. Even with this setback, images from the Hubble Space Telescope are up to ten times clearer than those that can be received on Earth.

Pictures taken by the Hubble Space Telescope (below) are clearer than any that can be obtained from Earth. A shows a picture of a galaxy taken by the Hubble Space Telescope; B is an enlargement; and D shows the picture after it has been cleaned up by a computer. C is the same galaxy as seen from a telescope on Earth.

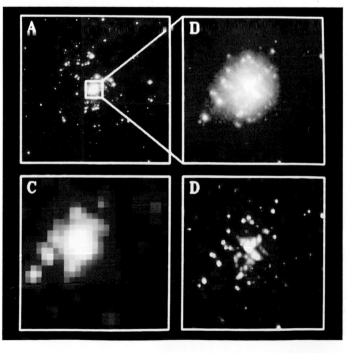

Invisible astronomy

Two instruments on board the Hubble Space Telescope are able to detect and analyze ultraviolet radiation that is invisible to the human eye. A series of probes planned for the 1990s will investigate this and other invisible forms of radiation. Some infrared radiation from deep space reaches Earth's surface, but most is absorbed by carbon dioxide gas and water vapor in the atmosphere. The first infrared satellite, the Infrared Astronomy Satellite (IRAS) launched in 1983, could see 100 times more than an Earth-based instrument. The latest infrared satellite, the European Infrared Space Observatory (ISO) will see 1,000 times more than IRAS. Whereas IRAS scanned and mapped the whole sky, ISO will be more selective. It will point at specially chosen objects for more detailed examination.

ISO will be followed by other missions that will form ESA's Horizon 2000 Space program. They include the Far Infrared and Sub-millimeter Space Telescope (FIRST), the X-ray Multi-Mirror telescope (XMM), and the Solar and Heliospheric Observatory (SOHO) (see page 25). XMM's three telescopes will give it the greatest X-ray collecting power of any mission so far. When it is launched in 1998, it will also be ESA's largest scientific satellite.

NASA's largest scientific satellite, the 15-ton Gamma Ray Observatory (GRO) was launched in April 1991. Gamma rays are produced when violent events in the universe, such as exploding or collapsing stars, produce very fast-moving particles. These particles punch their way into nearby atoms, and the result is a very high-energy wave called a gamma ray. Despite their high energy, gamma rays are absorbed by Earth's atmosphere, so gamma ray astronomers have to use instruments that are placed above the atmosphere. GRO may provide the first evidence of the existence of black holes — the collapsed cores of giant stars whose gravity is so strong that even light cannot escape.

The moon and Mars

The next destinations for manned space exploration in the solar system will be the moon and Mars. They are the two best candidates because conditions on all the other planets are too hostile to allow exploration and landings by people.

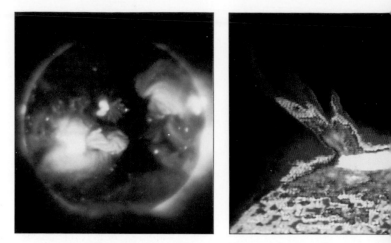

Ultraviolet and X-ray images of the sun. These images show how astronomy uses different types of radiation to reveal new information. The X-ray picture (left) shows activity high in the sun's atmosphere, while the ultraviolet image (right) shows activity deeper down, just above the surface.

Mining the moon

Twelve American Apollo astronauts have already visited the moon, but their missions were brief, and their rockets and spacecraft were not reusable. This time, if astronauts return to the moon, it will be to set up a permanent moonbase. The base would be used for scientific research and possibly commercial mining. The moon is rich in minerals, but for the foreseeable future it will not be worth the expense of bringing minerals all the way back from the moon. The most valuable material on the moon may turn out to be helium-3.

The first color picture of the surface of Mars taken by *Viking 1*, showing the fine reddish Martian soil.

Scientists are trying to develop a new type of power generator called a nuclear fusion reactor. Instead of splitting apart dangerous materials, such as uranium and plutonium, as existing nuclear reactors do, fusion reactors join safe materials, such as hydrogen and helium, to release the energy stored inside them. A special type of helium called helium-3 is a particularly good fuel for this. Although there is little helium-3 on Earth, it exists in large quantities on the moon. There could be enough helium-3 on the moon to provide all the electrical power requirements on Earth for many thousands of years to come.

It may be possible to begin building a moonbase by the year 2010. Its crews and supplies would be ferried from Earth to the space station *Freedom* (see pages 34 – 35) by spaceplane. A transfer vehicle would then transport the crews and their supplies to **lunar** orbit where another shuttle would take them down to the moon's surface.

Mars missions

Over the years, many different ways of sending astronauts to Mars and setting up a base have been investigated. Recent NASA studies suggest that the first phase of such a plan would involve four missions. The first would land a crew of three on Phobos, one of the Martian moons, and send an empty living module to Mars to act as backup living quarters for the second mission. Robots would also collect Martian soil samples and bring them to Phobos for analysis.

The second mission would land a crew of five on Mars, where they would stay for a year carrying out scientific research and exploration. The third mission, with a crew of five, would go to Mars and Phobos where it would build the first half of a system that would make rocket fuel from the soil and rocks. The fourth mission, also with a crew of five, would carry the second half of the rocket fuel system.

The first phase of four missions would prepare the way for further missions that would construct the base itself. The plan shows the first mission taking off in the year 2004. Each mission would last for up to three years, with a year between each crew returning from Mars and the next crew taking off.

radioactive – giving off energy in the form of particles and rays when the nucleus of an unstable substance, such as plutonium, radium, or uranium, breaks apart.
NASA – National Aeronautics and Space Administration. The main aviation and space research organization of the United States.
asteroids – rocky objects that orbit the sun, mainly between the orbits of Mars and Jupiter. They range in size from a few feet to hundreds of miles across.
magnetic field – the area around the poles of a magnet, in which the magnet can exert a force.
lunar – relating to, or describing the moon.

Space Survival

Throughout history, people have always explored the world around them. Each generation has sought out new places where people have never gone before. Space technology has enabled human beings to leave their home planet for the first time. A great deal of exploration and scientific research can be carried out by computerized space probes, but human beings are still the most versatile explorers, observers, and experimenters.

Until the NASA shuttle began operating in 1981, almost all astronauts were military pilots who were trained to do everything necessary during a mission. Space shuttle crews are made up of different types of astronauts trained to do different jobs. The commander and pilot fly the space shuttle. The mission specialist looks after the systems that link the orbiter craft with its payload. There may also be one or more payload specialists. These are scientists or engineers trained to

Construction in space. Erecting a tower from the cargo bay of the space shuttle *Atlantis*.

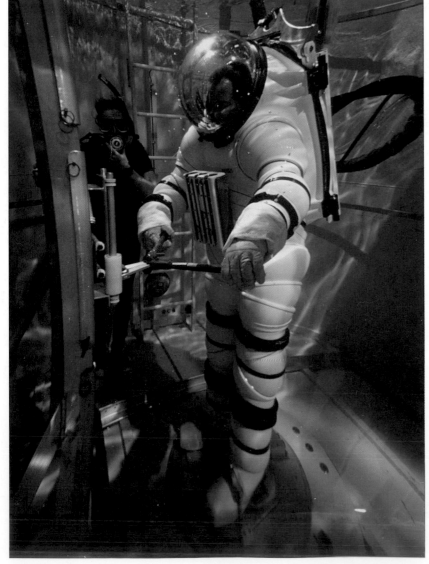

Spacesuit of the future. The AX-5 spacesuit being tested for comfort and flexibility. The AX-5 is one of the designs being considered for use on the space station *Freedom*.

INSIDE A SPACESUIT

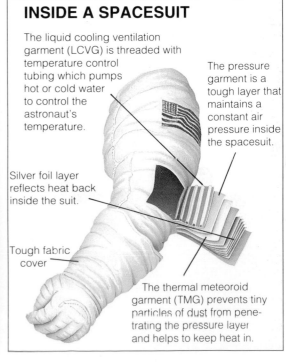

The liquid cooling ventilation garment (LCVG) is threaded with temperature control tubing which pumps hot or cold water to control the astronaut's temperature.

The pressure garment is a tough layer that maintains a constant air pressure inside the spacesuit.

Silver foil layer reflects heat back inside the suit.

Tough fabric cover

The thermal meteoroid garment (TMG) prevents tiny particles of dust from penetrating the pressure layer and helps to keep heat in.

operate the payload experiments. In 1991, the NASA space shuttle *Columbia* took off carrying three specialists, 29 rats, and thousands of jellyfish, in order to study the effects of weightlessness on organisms in space. In the future, there will be more specialists in, for example, construction, safety, and medicine.

Life-support systems

People cannot survive outside Earth's atmosphere. When they leave Earth, they must take an artificial atmosphere with them. The spacecraft's life-support system provides comfortable living and working conditions. The shuttle, for example, has a sophisticated environmental control and life-support system. Its many vital functions include recycling the spacecraft's air supply and adding fresh oxygen when necessary, keeping the air pressure the same as Earth's surface, and heating or cooling the air as necessary.

Sometimes astronauts have to go outside their spacecraft. On each Apollo mission, two astronauts ventured out onto the moon's surface. They each wore a portable life-support system carried in a backpack. Future astronauts will not only have backpacks for working in space, but also personal spacecraft for maneuvering in space and vehicles for moving around on the surface of other planets.

A personal spacecraft

If astronauts have to maneuver around outside the space shuttle orbiter, they can use a personal spacecraft called a Manned Maneuvering Unit (MMU). Astronauts, wearing a life-support backpack, simply clip their spacesuit to the MMU and fly away. Power for flight is provided by 24 nitrogen gas jets controlled from the end of the armrests. The MMU, or something like it, will be needed in the future by astronauts building and repairing large structures in space.

In 1984, Bruce McCandless became the first human to fly in space without any link to a spacecraft. Using a Manned Maneuvering Unit (MMU) he flew more than 300 feet away from the space shuttle.

Reusable spaceplanes

Until 1981, all satellites and deep space probes were launched by rockets that were used only once. If space travel and exploration are to develop and flourish in the future, with crews making regular visits to laboratories and manufacturing systems in orbit, then a new type of spacecraft that can be used over and over again has to be developed. The NASA shuttle is the first step in developing such a vehicle.

The shuttle

The NASA shuttle was launched for the first time on April 12, 1981. It consists of a reusable orbiter craft, in which the crew lives and works, connected to a fuel tank and two booster rockets. The orbiter's three main engines and both boosters are needed to lift the shuttle off the launch pad and begin its ascent into orbit. When their job is done, the boosters and fuel tank are dropped into the Atlantic Ocean. The tank is destroyed but the boosters survive. They are collected by ship and refueled for later use.

The NASP plane. It is shown here about to rendezvous with a satellite in Earth orbit.

A fleet of three shuttle orbiters, called *Columbia*, *Atlantis*, and *Discovery*, now make regular trips to Earth orbit and back again. A fourth orbiter, *Challenger*, was destroyed by an explosion just after its launch on January 28, 1986, killing its crew of seven.

The next step is to build a completely reusable space vehicle. But that is still some way off in the future. In the meantime, several countries have announced plans to build shuttle craft. The only other country to have launched a space shuttle is the former Soviet Union. Its *Buran* (meaning Snowstorm) space shuttle has made one crewless test flight.

The next step — NASP

The most ambitious spaceplane project is the United States' National Aero-Space Plane (NASP). The project to build it is code-named X-30. NASP will take off from a normal airport runway, accelerate to 25 times the speed of sound (1,000 feet per second in air) and either ascend to orbit the Earth or descend again to land at another airport. It will be able to fly from Washington, D.C. to Tokyo, Japan, within two hours instead of the sixteen hours taken

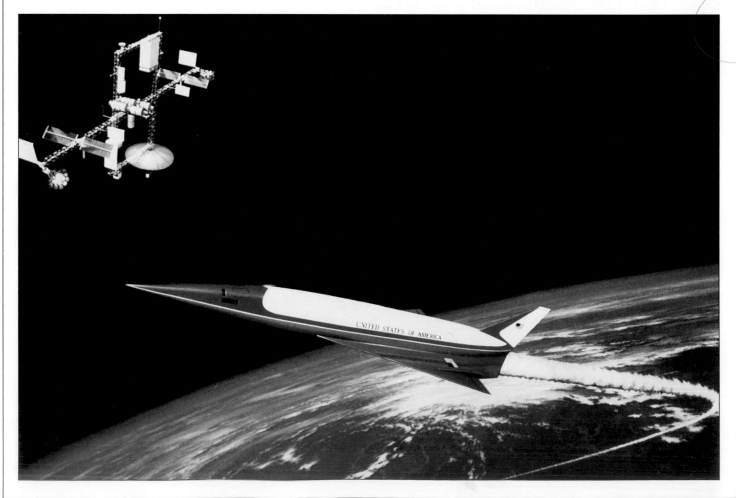

HOTOL (above) taking off from the back of a Russian Antonov AN-225 aircraft at a height of 30,000 feet. *Sänger* (left) is shown here in a one-eighth scale model.

by aircraft today. The X-30 project will require new materials to withstand the immense temperatures caused by air rubbing against the outer skin of the plane, as well as a new type of engine that can operate at such high speeds.

Hermes
Meanwhile in Europe, the European Space Agency (ESA) is designing its own space shuttle called *Hermes*. The European shuttle, which will be launched on top of the European *Ariane 5* rocket, will be smaller than the NASA shuttle. Heavy payloads will be launched separately by Ariane rockets. *Hermes* will be able to carry a crew of three. Flights are due to begin at the turn of the century when *Hermes* will be used to fly to the American space station, *Freedom* (see pages 34 – 35).

HOTOL
In Britain, a spaceplane called HOTOL (Horizontal Take Off and Landing) is being developed as a crewless satellite launcher. HOTOL will use a revolutionary new engine being developed by Rolls-Royce that will operate like a jet engine while HOTOL is flying in the atmosphere and like a rocket engine when HOTOL is in space. Originally designed to take off from the ground, HOTOL may now be launched from the back of the world's largest and heaviest aircraft — the Russian-built Antonov An-225.

Sänger
Other countries including Germany and Japan are also planning to build spaceplanes. The German design is a two-stage craft called *Sänger*. A jet-powered mother craft carries the orbiter, called *Horus*, on its back. After takeoff from a runway, the mother craft will accelerate to more than six times the speed of sound. Then, at an altitude of about 21 miles, the rocket-powered *Horus* spaceplane will separate from the mother craft and continue into orbit.

33

Living in Space

Not every experiment, observation, or manu-facturing process fits into the normal space shuttle mission's short duration of a few days. Many of these activities are better suited to much longer missions. The next logical step in space research is, therefore, the development of a craft with crew kept permanently in orbit around Earth — a space station. Space stations may also act as staging posts where spacecraft change crews and take on supplies and fuel before embarking on longer flights, to the moon for example. Spacecraft for the longest manned missions, to the planet Mars and be-yond, may be built by construction workers based in a nearby space station.

The United States and the former Soviet Union have both experimented with small space stations. The United States operated its Skylab space station in the 1970s, but by the early 1990s the former Soviet Union had gained the most experience in space station development and operation with its Salyut and larger Mir craft. The European Space Agency (ESA) has built an orbiting laboratory called Spacelab that fits inside the Space shuttle's payload bay. It has given European astronauts valuable experience working in space.

Freedom

In 1980, President Ronald Reagan committed the United States to building a permanently manned space station. Construction of the first phase of the station, which will be known as *Freedom,* will begin in the mid-1990s. The first temporary crews of scientists could be work-ing on board *Freedom* by 1997, and it should have permanent crews by early next century. The first phase will involve constructing a 500 foot-long beam 200 miles above Earth. Living quarters and working areas, called modules, will then be attached to the beam. Each mod-ule will be about 40 feet long. A docking port for the space shuttle will also be fitted. Electri-cal power will be provided by huge solar pan-els, each measuring almost 100 feet by 30 feet. In addition to the American modules, both Europe and Japan will supply modules to attach to the station.

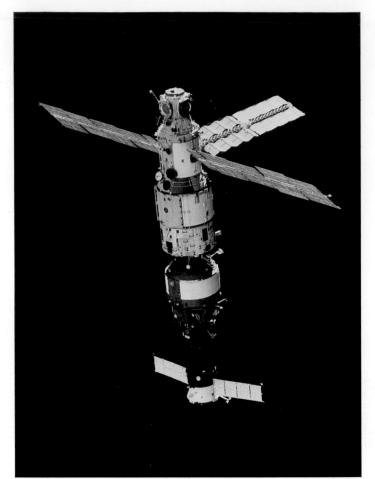

The Mir space station (with three solar panels) in orbit, docked with a Soyuz spacecraft.

What is a space station for?

The *Freedom* space station will orbit about 200 miles above Earth. Its crew will have various jobs including checking satellites before they are launched, repairing satellites, performing experiments in the laboratory, and monitoring Earth. The station might be used as a fuel base to refuel spacecraft traveling farther afield — to the moon or to Mars. It will also provide an opportunity to study the effects of weightless-ness upon the human body, vital if humans are to stay in space for any length of time. And one day in the future it might also mean that ordi-nary people will get the chance to visit and ex-perience life in space for themselves.

Columbus

The United States invited other countries and space agencies to take part in the *Freedom* space station project. The European contribution, called *Columbus,* will include one of the mod-ules at the heart of the station. Between four and eight astronauts will live and work there

The *Freedom* space station.
In this artist's impression a space shuttle is seen docked with the station.

LIVING IN A SPACE STATION

A habitation module on the *Freedom* space station will be fitted out with everything a crew needs to survive in space, including an environmental life-support system that provides a breathable atmosphere and water. Important or dangerous areas, such as the hatches and handrails, will be highlighted for the crew by painting them in bright colors.

wardroom where crews will eat, work, and relax.

Earth observation window

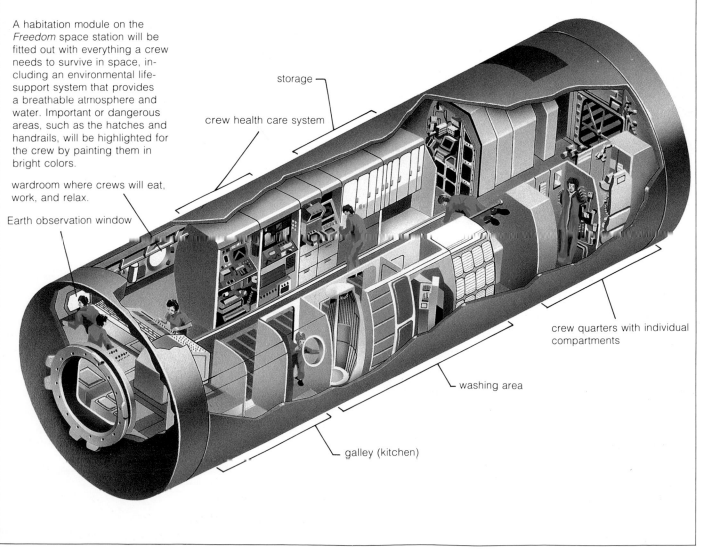

storage

crew health care system

crew quarters with individual compartments

washing area

galley (kitchen)

in a laboratory attached to the module. The *Columbus* project will also provide two orbiting platforms that will not be attached to the space station. The *Columbus* Free-Flying Laboratory will orbit the Earth near the space station. It will carry experiments that are too delicate to be carried out on the main space station, where they could be affected by changes in altitude or the vibrations caused by astronauts moving around. Astronauts will visit it from time to time to change the experiments and equipment. The *Columbus* Polar Platform will orbit Earth from pole to pole collecting information to help study the Earth's climate.

Manufacturing in space

Earth's gravity is essential for life on Earth, but it can present serious problems for the manufacturers of some materials and products. On Earth, liquid mixtures often suffer from an effect called sedimentation. The largest particles in the mixture are pulled downward, leaving the smallest particles at the top. When the mixture sets into a solid, it does so unevenly. In orbit around Earth, where there is no gravity,

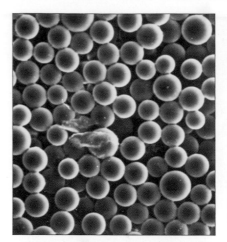

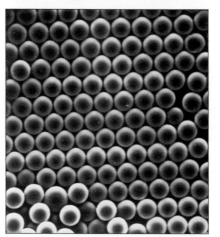

The microspheres in the bottom picture were made during four space shuttle flights between 1982 and 1984. Similar spheres made on Earth (top) are distorted by gravity. The spheres, used for calibrating instruments and measuring the size of microscopic particles, were the first products made in space to go on sale on Earth.

A **Protein Crystal Growth** (PCG) grown in space as part of NASA's program for the commercial development of space.

European astronaut Ulf Merbold (left) operates a furnace used for metal processing and crystal growth.

mixtures remain perfectly mixed and set evenly. Convection is another problem. If all of a liquid is not at the same temperature, the warmer regions tend to float up toward the top, and the cooler regions sink to the bottom. Like sedimentation, convection does not occur in space.

Making medicines
The manufacture of modern medicines often involves the production of extremely pure substances. Sedimentation and convection prevent the purest substances from being made. Much purer materials for medicines can be made in space. The cost of making them in space is enormous, but there are some substances that are worth it because of the increased quantities of much purer samples that can be produced. These space-manufactured medicines may, one day, help to provide cures for many illnesses.

Crystals, glass, and metals
Large pure crystals are used in many areas of technology, from computer chips to lasers. Gravity causes crystals to develop faults as they grow. In space they can grow unhindered by gravity. Large, pure crystals have already been made in space where they grew to 50 times the size of those produced on Earth. In the future, larger quantities of many more crystals will be made in space for industries on Earth.

Some materials are difficult to make because they react with their containers and become impure. These can be made very purely in orbit because, without gravity, they do not need any containers. They can be held and moved around by jets of inert gas or, if they can be magnetized, they can be held in a magnetic field called a magnetic bottle. Glass manufactured on Earth is often contaminated by the container in which it is made. In space, extremely pure glass for use in lasers and optical systems, such as telescopes, can be made by containerless processing.

Some materials produced at very high temperatures are well-suited to space manufacturing. In a process called sintering, metal powders are converted to solids by heat and pressure. Parts of jet engines, motors, and a variety of magnetic materials are made by sintering. In space, the end result is more evenly mixed and stronger than the Earth-made product.

At first, only small quantities of very expensive materials are likely to be made in space. Later, when larger manufacturing facilities are available in space stations and free-flying platforms, the gravity-free conditions in space will be used for making larger quantities of many more materials.

Dangers in space
In many ways, space is a much friendlier place for scientific equipment than Earth. In space, there is no water to short-circuit electrical equipment, no clouds to obscure the view, no windblown sand and dust to scratch surfaces or settle on lenses, and no surface vibrations to shake equipment. However, beyond the protective blanket of Earth's atmosphere, space presents some new dangers of its own to both equipment and space travelers. There are two major dangers — space junk and radiation.

Space junk

Space junk, or debris, ranges in size from microscopic particles to pieces of rockets and satellites several feet across. The larger metal objects, from about 4 inches across, can be tracked by radar or seen by optical instruments on Earth.

But without the protection of Earth's atmosphere, small particles that cannot be tracked are dangerous too. During the seventh NASA space shuttle flight, something struck the spacecraft's window and left a crater in the glass about the size of a small pebble. When it was examined, traces of titanium oxide, aluminum, carbon, and potassium were found. These are the ingredients of white paint! The shuttle had been struck by a tiny flake of paint measuring only 0.01 inches across. Perhaps it was from a satellite.

The larger a spacecraft is, and the longer it stays in orbit, the more chance it has of being struck by something. Space stations have the greatest chance of being hit. The *Freedom* space station will probably be protected by plastic shields or bumpers.

Radiation

Space radiation may be in the form of particles or radio waves from the sun and from distant stars and galaxies. High doses of radiation, mainly from the sun, can cause illness and in extreme cases, death. Sudden bursts of activity on the sun's surface can cause an equally sudden increase in radiation reaching Earth. The atmosphere blocks most of it, but astronauts and even pilots of high-altitude aircraft, such as the Concorde, can be affected. Astronauts serving on board space stations and on long-duration missions to the planets will have to be protected, possibly by building storm shelters to which they could retreat during periods of intense radiation.

The physical effects

In addition to radiation, astronauts face a number of possibly harmful physical effects on their bodies if they stay in space for long periods. The human body evolved for life in Earth's atmosphere and gravity. In space, without Earth's gravity for the body to work against, the body begins to change, or adapt,

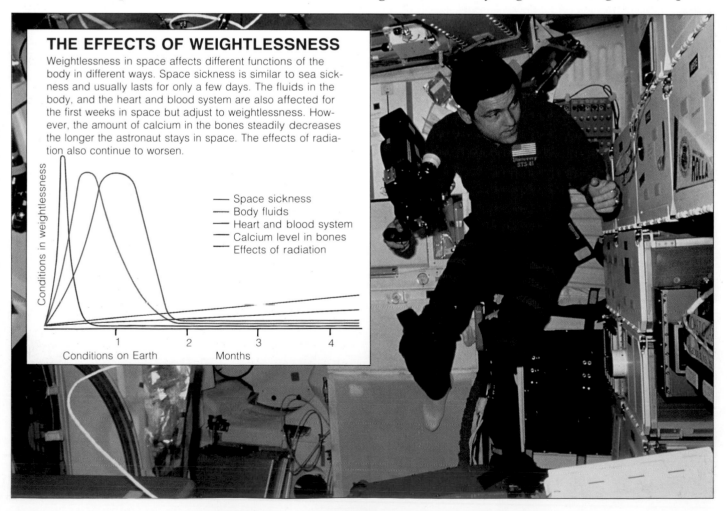

THE EFFECTS OF WEIGHTLESSNESS

Weightlessness in space affects different functions of the body in different ways. Space sickness is similar to sea sickness and usually lasts for only a few days. The fluids in the body, and the heart and blood system are also affected for the first weeks in space but adjust to weightlessness. However, the amount of calcium in the bones steadily decreases the longer the astronaut stays in space. The effects of radiation also continue to worsen.

Conditions in weightlessness

— Space sickness
— Body fluids
— Heart and blood system
— Calcium level in bones
— Effects of radiation

1   2   3   4
Conditions on Earth   Months

Astronauts are continually tested and monitored to investigate how they respond to living in space (below). Exercise is important — cosmonaut Yuri Romanenko is shown here on a running machine in the Mir space station (above). It is also vital that crews can amicably live and work together (top right).

to its new conditions. Some changes, such as space sickness, are usually temporary. Without gravity, some astronauts feel dizzy and sick. The illness is similar to seasickness on Earth and usually lasts no longer than a few days.

Other changes are more serious. Muscles begin to become thinner because of the lack of exertion against gravity, and bones begin to lose the calcium that gives them their strength. The problems caused by these and other effects can last long after the astronaut returns to Earth. Cosmonauts from the former Soviet Union have been very successful in minimizing these physical effects by using a special program of hard exercise. Some cosmonauts have stayed in orbit for a year. However, exercise may not be enough to keep astronauts healthy for a longer journey, to Mars for example, lasting perhaps three years. The spacecraft may have to be spun to produce an artificial gravity. To investigate this, it is likely that an artificial gravity test chamber will be built on the *Freedom* space station, or on a free-flying experimental platform orbiting in space nearby.

Psychological problems
Up to now, most space missions have been short and full of activity. Astronauts are usually kept so busy that they do not have a chance to get bored. Longer stays on board space stations or on flights to Mars or to the outer planets could involve long periods with little activity. Loneliness and boredom could cause serious mental problems, which spacecraft designers and psychologists will have to join forces in order to solve.

Space Colonies

As far as we know, Earth is the only planet in the universe where intelligent beings live. After millions of years of evolution, human beings have now begun to move out from their home planet. At some time in the future, astronauts and their families are likely to be based permanently in space stations or settlements on the moon. Like the first Europeans who went to live in the new worlds of Australia and North America, they will take an initial payload of supplies with them, but after that they will have to make or grow everything they need.

The inhabitants of such a colony or settlement would have to be protected from hazards such as fire, radiation from space, pests or diseases in the food, leaks or other failures in the structure, and contamination of the water supply. On Earth, if conditions in one place become intolerable, the people affected can often go elsewhere. However, in a space colony or a moon settlement, there is nowhere else to go. The structure must be safe, and the living system of people and plants must survive together.

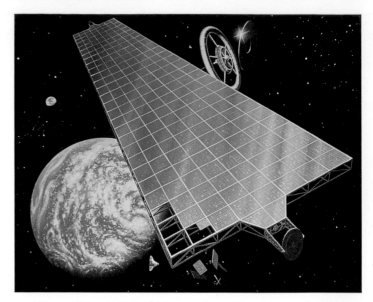

A solar satellite power station being constructed in space. It could generate electric power from sunlight and transmit it to Earth. In this artist's impression, the wheel-shaped colony where the construction workers live can be seen in the background.

Terraforming

The greenhouse effect and damage to the ozone layer has taught us that human activities can affect a planet's climate. In theory at least, it may be possible to change the climate of another planet intentionally. Instead of polluting and damaging the planet's atmosphere as we have

MOVING TO MARS

1. The first astronauts arrive on Mars and set up a manned base inside biospheres. They grow food plants in the artificial atmosphere of these protective domes. Terraforming begins as machines are built to pump out greenhouse gases into the Martian atmosphere.

2. As clouds and water begin to form, the Martian sky starts to change from pink to blue. There is more oxygen in the atmosphere, but rebreathers (face masks attached to small backpacks) still have to be worn. More astronauts arrive and cities begin to develop.

done on Earth, some scientists believe that the atmosphere of another nearby planet might be changed from one that will not support human life to one that will. Creating Earth-like conditions on other worlds is called terraforming, from the Latin word *terra* meaning earth.

The task would be huge, very costly, and it would take centuries to accomplish. Mars and Venus are the prime candidates for terraforming. Conditions on Venus are thought to be so extreme that it is probably beyond the scope of terraformers. Its average surface temperature is over 800°F, hot enough to melt lead. But some scientists believe that the Martian atmosphere could be changed to create a more Earth-like world. One plan involves converting materials already present in the Martian surface into gases that have a similar heating effect to carbon dioxide. They would begin to trap more of the sun's energy and raise the planet's temperature. Ice at the Martian poles reflects heat back into space, so the polar ice might be darkened by encouraging simple plants to grow over it, or by melting it using huge mirrors in space to direct sunlight onto the poles.

As the ice melted, and the water evaporated and rose into the warming atmosphere, clouds would begin to form, and the Martian sky would change from its present pink color to a more Earth-like blue. Rain would fall on the planet for the first time in millions of years, lakes would form, and rivers would flow. The amount of carbon dioxide in the atmosphere would fall as it dissolved into the surface water, and plants would begin to spread across the planet. The whole process would take anything from a few centuries to several thousand years. It seems a very long time, but compared to the time scale of a planet's natural evolution, which normally takes thousands of millions of years, it's a very short time indeed.

No attempt to begin terraforming Mars is likely until the first people land on Mars to study the planet. That is expected to happen during the middle of the next century.

Space for thought
● Imagine a leisure resort in space. Think how the absence of gravity might affect the games we play on Earth. What new games might be possible in space that couldn't be played on Earth?

3. The temperature of Mars is still rising and the whole planet begins to turn green with vegetation. Rivers and seas form and crops can be harvested. But there are still no non-human animals because there is not enough oxygen.

4. Oxygen made by machines is pumped into the atmosphere. Rebreathers are unnecessary and non-human animals can be introduced.

3

4

Conclusions

Some trends in space science are clear. Scientific satellites will probably become smaller and more numerous if the size of their on-board equipment allows; while communications satellites and space stations seem certain to become even bigger. Their size is only limited by the capacity of the rockets needed to launch them. Spaceflight will continue, with astronauts living on space stations, lunar bases, and perhaps an astronaut mission to Mars early next century.

However, no branch of science or technology exists in isolation. Discoveries and developments in one branch are often taken up and used by scientists and engineers working in completely different fields. Space science is no exception, and these spin-offs will continue to enable everyone to benefit from new developments in space research.

Many thousands of new materials and pieces of equipment developed for space systems have already been applied elsewhere. Modern protective suits and breathing apparatus used by firefighters are adapted from space systems developed in the 1970s. Until then, the breathing equipment used by firefighters was so big that it could not be used in small spaces, and so heavy that it could not be carried for long before the firefighters were too tired to go on. The new system, developed at the Johnson Space Center in Houston, Texas, was one-third lighter and much smaller. Nowadays, every major manufacturer of breathing equipment is using space technology of some sort.

Cordless power tools, widely used in the home industry, were developed from tools designed to allow Apollo astronauts to collect samples of lunar soil. The equipment used to monitor patients in hospitals uses technology developed to monitor astronauts.

A hand-controller used by severely handicapped car drivers was developed from the joystick designed for the Apollo Lunar Rover. This four-wheeled electric car allowed Apollo astronauts to drive across the moon's surface using only one hand.

Medical scientists have developed many different ways of obtaining images from inside the human body. Space science has provided new ways of improving and analyzing the pictures. One particular technique, called Nuclear

The first woman to walk in open space (right), the Russian cosmonaut Svetlana Savitskaya, is seen here using a general-purpose hand-tool to weld metal plates. Cordless hand-tools (inset) have also been adapted for use on Earth.
Fire-fighting suits and monitoring machines in hospitals are two examples where Space technology has been adapted for use on Earth.

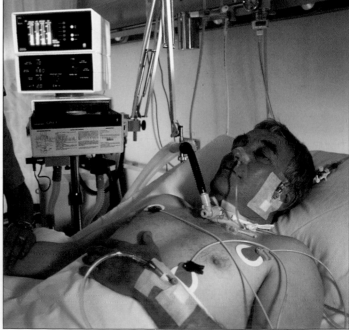

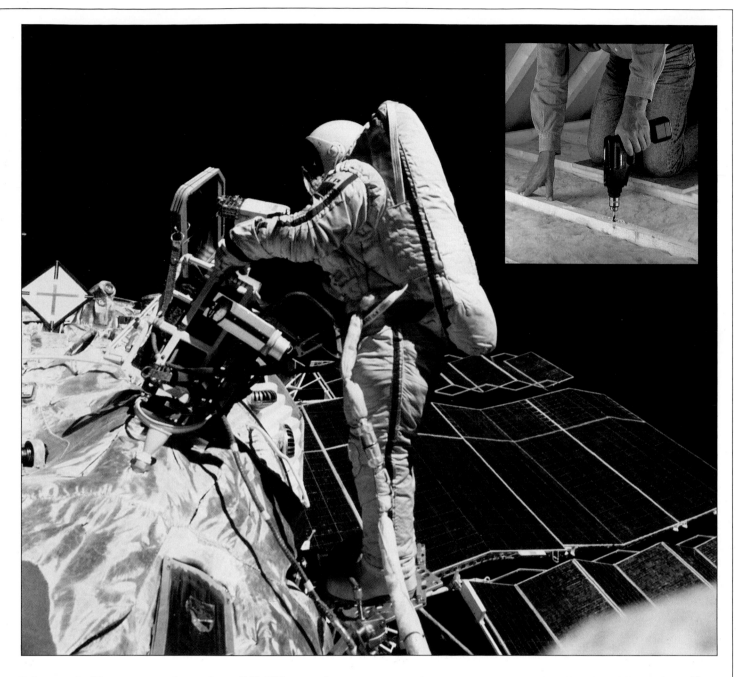

Magnetic Resonance imaging (NMR), produces images that are more difficult to interpret than X-ray pictures. Space scientists had already developed ways of converting the enormous amount of information sent back to Earth from distant space probes into easily understandable pictures of the planets and stars. These same methods were also applied to NMR images.

Cape Canaveral, the launch site from which most American rockets are launched, is on the east coast of Florida. Without protection from the constant windblown, salty, sea spray, the metal launch towers would have become dangerously weakened by corrosion. Special paints and other coatings developed to protect the towers were soon being used to protect bridges, pipelines, ships, and oil rigs.

Military space projects also lead to spin-offs in non-military technology. A weapon developed for the Strategic Defense Initiative (SDI) or "Star Wars" project is now used to treat cancer. Its "Star Wars" objective was to destroy enemy satellites in space with a powerful beam of particles called protons. But a proton beam can also be programmed to treat cancerous cells without damaging any of the healthy cells around the cancer.

In the near future, the *Freedom* space station and the next generation of spaceplanes and satellites are likely sources of new discoveries that may someday be applied to aircraft engines, new materials, food technology, power generating systems, medicine . . . and who knows what else!

Glossary

antenna – a metal wire, frame, plate, or dish used to detect radio signals.

asteroids – rocky objects that orbit the sun, mainly between the orbits of Mars and Jupiter. They range in size from a few feet to hundreds of miles across.

atmosphere – the thin layer of gases that surrounds Earth, composed mainly of nitrogen and oxygen.

black holes – the collapsed cores of giant stars. Their gravity is so strong that even light cannot escape.

electromagnetic waves – waves of energy composed of vibrating electric and magnetic fields. Light, radio, X-ray, infrared, and ultraviolet are all examples of electromagnetic waves.

electromagnets – magnets made from a coil of wire wound around an iron bar. When an electric current flows through the coil, the bar becomes a magnet.

equatorial – near the equator, an imaginary line around Earth's middle, midway between the poles.

facsimile transmission (fax) – a system for sending printed text, drawings, handwriting, and pictures by telephone through a fax machine.

frequency – the number of cycles or waves that pass any point in a second. The waves could be water waves, sound, light or radio waves.

friction – a force produced when things rub against each other. Friction between a rocket and the air rushing past it slows the rocket down.

galaxies – collections of millions upon millions of stars traveling through space together. Our star system is part of the galaxy called the Milky Way. Galaxies are separated by enormous distances.

gamma rays – given off by radioactive substances such as plutonium or uranium. They have very short wavelengths.

infrared – radiation with a longer wavelength than red light, therefore invisible to the human eye, but felt as heat.

invisible radiation – waves of energy, such as ultraviolet and X-ray, that are not visible to the human eye.

laser – a device that produces an intense beam of light.

light-year – the distance that light travels through space in one year.

lunar – related to or describing the moon.

magnetic field – the area around the poles of a magnet, in which the magnet can exert a force.

navigation – finding a way from one place to another and plotting the position upon Earth's surface.

orbit – the path followed by an object, such as a spacecraft or a moon, as it circles a planet, or the path of a planet around a star.

payload – the equipment carried by a rocket or satellite.

radar – (RAdio Detecting And Ranging) a system used to locate objects and draw maps of a planet's surface by sending out bursts of radio waves and analyzing any reflections that bounce back.

radiation – rays of energy and particles that reach Earth from outer space.

recycling – using something again instead of throwing it away.

sensors – the part of a satellite's payload that reacts to whatever it is designed to detect — heat, light, or radio energy, for example. This change is converted to an electrical signal and radioed to Earth.

solar system – the group of nine planets, including Earth, which orbit the star we call the sun.

solar wind – the invisible stream of particles that stream into space from the sun's surface.

sunspot – a dark spot on the sun's surface; a sunspot is slightly cooler than the surrounding material.

ultraviolet – invisible to the human eye, these waves usually come from the sun. They are mostly absorbed by Earth's atmosphere. Ultraviolet waves make you tan and can be dangerous, causing skin cancer and cataracts from long exposure.

universe – the whole of space, including everything in it, to infinity.

wavelength – the distance between two neighboring vibrations of any wave.

Index